The Little Book of Investigations

Science in the Foundation Stage

by Sally Featherstone

Illustrations by Martha Hardy

LITTLE BOOKS WITH BIG IDEAS

Reprinted 2009, 2010, 2013
Published 2009 by Featherstone,
An imprint of Bloomsbury Publishing Plc
50 Bedford Square, London, WC1B 3DP
www.bloomsbury.com

First published in the UK by Featherstone Education, 2003

ISBN 978-1-9041-8766-0

Series Editor, Sally Featherstone

Illustrated by Martha Hardy

A CIP record for this publication is available from the British Library.

Printed in Great Britain by Latimer Trend & Company Limited

This book is produced using paper that is made from wood grown in
managed, sustainable forests. It is natural, renewable and recyclable.
The logging and manufacturing processes conform to the environmental
regulations of the country of origin.

Contents

Introduction

This book is intended for practitioners working with young children in schools, playgroups, nurseries, informal groups and, of course, at home.

In **Knowledge and Understanding of the World**, the Curriculum Guidance for the Foundation Stage says:

- "To give all children the best opportunities for developing effectively their knowledge and understanding of the world, practitioners should give particular attention to:
  - activities based on first hand experiences that encourage exploration, observation, problem solving, prediction, critical thinking, decision making and discussion;
  - an environment with a wide range of activities indoors and out of doors that stimulate children's interest and curiosity;
  - adult support in helping children communicate and record orally and in other ways.

  They learn the skills necessary to this area of learning by using a range of tools; for example computers, magnifiers, gardening tools, scissors, hole punches and screwdrivers. They learn by encountering creatures, people, plants and objects in their natural environments and in real life situations, for example in the shop or the garden. They learn effectively by doing things, for example by using pulleys to lift heavy objects or observing the effect of increasing the incline of a slope on how fast a vehicle travels."

This book is intended to help you to provide and structure some of these experiences, using your skills to support children in their own learning, asking the right questions and intervening at just the right time to extend interest and involvement.

Each of the investigations starts with a period of free exploration, either in your setting, your garden, or the local community. The resources and equipment suggested are flexible and you can substitute those you haven't got with others which are more readily available. Most of the investigations use materials and equipment easily found or collected, and a reference page at the end of the book gives you addresses for more unusual things, such as the Butterfly Box (pages 48-49).

Many of the familiar activities planned for young children make a suitable starting point for an investigation – a deeper look at something. A walk, a bowl of water, a ladybird, a rainbow, a pot of glue – these and many more provide opportunities for you to help children to look more closely at familiar objects and events.

You can employ this book in several ways. The materials can be used to:

- extend opportunities that arise naturally in your setting – using informal starting points, and referring to the book for extensions and further ideas;
- plan practical activities with a purpose – starting children off in new or adapted versions of familiar play situations, then taking the activity further by adult observation and interaction;
- plan adult directed activities for observation and assessment purposes – giving the children plenty of time to explore the materials before you begin to observe and record their learning.

Each activity meets a range of the Early Learning Goals for the Foundation Stage, and you will find some of these identified on each page. You can use these goals to help with your planning or with observation for profiles.

The key early learning goals explored in this book are:

In **personal, social and emotional development**:

- continue to be interested, excited and motivated to learn;
- be confident to try new activities, initiate ideas and speak in a familiar group;
- select and use activities and resources independently;
- work as part of a group or class, taking turns and sharing fairly, understanding that there need to be agreed values, and codes of behaviour for groups of people, including adults and children, to work together.

In **communication, language and literacy**:

- interact with others, negotiating plans and activities and taking turns in conversations;
- extend their vocabulary, exploring the meanings and sounds of new words.

In **knowledge and understanding of the world**:

- investigate objects and materials by using all of their senses as appropriate;
- find out about, and identify some features of living things, objects and events they observe;
- look closely at similarities, differences, patterns and change;
- ask questions about why things happen and how things work;
- build and construct with a wide range of objects, selecting appropriate resources, and adapting their work where necessary;
- select tools and techniques they need to shape, assemble and join the materials they are using;
- observe, find out and identify features in the place they live and the natural world;
- find out about their environment, and talk about those features they like and dislike.

In **physical development**:

- use a range of small and large equipment;
- handle tools, objects, construction and malleable materials safely and with increasing control.

In **creative development**:

- explore colour, texture, shape, form and space in two and three dimensions;
- respond in a variety of ways to what they see, hear, smell, touch and feel;
- express and communicate their ideas, thoughts and feelings by using a widening range of materials, suitable tools, imaginative and role-play, movement, designing and making, and a variety of songs and instruments;
- recognise and explore how sounds can be changed, sing simple songs from memory, recognise repeated sounds and sound patterns and match movements to music.

In recording their investigations, children will also be:

- using their phonic knowledge to write simple regular words and make phonetically plausible attempts at more complex words;
- attempting writing for various purposes, using features of different forms such as lists, stories, instructions.

Encouraging Independence and Thinking Skills

Some of the activities in this Little Book require adult help with preparation, and in some cases help will be needed with the activity itself. Younger children, or those with special needs, will need more support, but we would encourage you to let children do everything they can themselves, working together only when needed.

Even the youngest children can:

- prepare, organise and protect their own work area
- collect objects for the investigation (in the setting or at home)
- choose the materials and equipment for their own work
- explore the materials you provide
- stick, spread, cover, fill and mix
- record in pictures, photos, words, and 'have a go' writing
- suggest extensions to observations
- look carefully and say something of what they see
- help to clear up and put away equipment

While you concentrate on:

- watching, listening and helping when they ask you
- noting how children learn and what they know and can do
- offering additional resources
- modelling the skills you want them to use
- explaining the use of new or unfamiliar equipment or skills
- accompanying them in their investigations

Asking open questions

Open questions give children opportunities to:

explain: "How did you do that?" "What happened next?"

demonstrate: "Can you show me how?"

predict: "What do you think will happen?" "What could you do next?

think: "Why do you think.......?" "How did that happen?"

solve problems: "How could we do that?"
"How could we make that work better?"
"What could we use to make that work?"

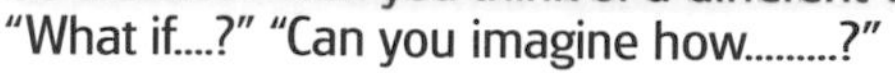

be creative: "Can you think of a different way to ....?"
"What if....?" "Can you imagine how.........?"

Investigate dissolving

Explore things that disappear when you put them in water.

What you need:

I will need

- warm water
- white and brown sugar, salt and sand
- saucers or small bowls
- plastic containers or small jars (with screw tops if possible)
- teaspoons
- magnifying glasses

Key words

- stir
- mix
- add
- disappear
- dissolve
- grains
- lumps
- bottom
- jar

Early Learning Goals

CLL – extend their vocabulary, exploring the meanings and sounds of new words

PSED – continue to be interested, excited and motivated to learn
– select and use activities and resources independently

KUW – investigate objects and materials by using all of their senses as appropriate
– look closely at differences and change
– ask questions about why things happen and how things work
– observe, find out and identify features in the natural world.

1. Explore

- Put some salt, sugar and sand in saucers or bowls.
- Let the children explore the substances. Touch, smell and taste them. (Tasting sand is not recommended!)

2. Talk

- Talk with the children about the differences between the different substances.
- Use the magnifying glasses to see differences and explore the grains of sand and sugar.

3. Experiment and observe

- Now help the children to pour some warm water into several jars or pots.
- Help them to carefully spoon some sugar, sand and salt into three different jars.
- Stir the water and watch what happens. Stop stirring frequently to talk about what is happening. Use the new words as you describe what is happening.
- Talk about where the salt and sugar have gone and why the sand doesn't disappear.

4. Record

- Take a series of photos of the experiments.
- Record what the children say, including use of new vocabulary.
- Draw some pictures of jars with a pencil or fine felt pen. Then let the children add 'before' and 'after' images.

5. Taking it further

- Talk to the children about whether you can get the salt or sugar back out of the water. Try some of their ideas.
- Expand your investigation of dissolving using flour, soap, cellulose paste, mud, white glue, powder paint. Discuss what happens to each substance.

Investigate melting

This investigation can use the sun or an artificial heat source to experiment with melting indoors or out.

What you need:

- chocolate
- jelly, ice cubes, butter, and soft 'jelly' sweets
- teaspoons
- hairdryer or hot water
- plastic bowl
- foil or greaseproof paper

Key words

- smelt
- change
- runny
- liquid
- soft
- hard
- careful
- watch
- touch
- heat
- cool
- taste

Early Learning Goals

PSED – select and use activities and resources independently

KUW – investigate objects and materials by using all of their senses as appropriate

– find out about, and identify some features of living things, objects and events they observe

– look closely at differences and change

– ask questions about why things happen

– observe, find out and identify features in the natural world

CD – respond in a variety of ways to what they see, hear, smell, touch and feel

1. Explore

- Put some chocolate, jelly, butter, or a sweet on the path or playground in the sun.
- Watch what happens. Poke the stuff with a stick or finger to see what it feels like. Watch it drizzle, drip, spread across the ground.

2. Talk

- Talk with the children about what happens when things get hot. Discuss things that melt and things that don't.
- Watch again to see what happens as the material cools.

3. Experiment and observe

- Now help the children to break some chocolate into a plastic bowl and either heat it with a hair dryer, or stand the bowl in hot water (take care!).
- Watch as the chocolate melts, stop frequently to discuss what is happening.
- When the chocolate has melted, spoon small amounts onto foil or greaseproof paper and watch again as the chocolate cools and hardens into chocolate drops.

4. Record

- Take some photos. A series of chocolate melting could make a sequencing game.
- Record what the children say, including new vocabulary.
- Draw pictures with pencil or fine felt pen.
- Make a picture list of things that melt.

5. Taking it further

- Investigate ice melting. Freeze (or buy) lots of ice cubes and put them in a water tray. Watch what happens as you play with them.
- Melt butter or margarine and make flapjacks.
- Make jelly. Watch the jelly cubes melt and then solidify again.
- Melt chocolate. Pour it over ice cream. Watch it cool and harden.

Investigate mixing

Look carefully at how things mix together, and have some fun with the results.

What you need:

- 4 or more cups of cornflour
- 2 cups of water
- a large tray, bowl or water tray
- food colouring
- aprons
- floor covering

Key words

- mix
- pour
- change
- wet/dry
- drip
- stringy
- pattern
- powder
- lift
- fingers

Early Learning Goals

PSED – be confident to try new activities
– select and use activities independently

KUW – investigate objects and materials by using all of their senses as appropriate;
– ask questions about why things happen and how things work

PD – use a range of small and large equipment
– handle malleable materials safely and with increasing control

CD – respond in a variety of ways to what they see, hear, smell, touch and feel

1. Explore

- Put the cornflour in the tray.
- Let the children explore the dry powder.
- Now add some water and help the children to mix the water in with their hands.

2. Talk

- Talk with the children about what happens as the water and cornflour mix. Talk about and ask questions about change.
- Encourage them to describe what the goop feels like.

3. Experiment and observe

- Watch as the children play with the mixture.
- Let them tell you what it feels like and how it behaves.
- Look carefully at what happens when the mixture is left to run together in the tray. Compare this with the way it behaves when you pick it up in your hands. Encourage careful looking.
- Model the key words by using them as you play.
- Add some food colouring and watch what happens.

4. Record

- Make some notes of how individuals respond, what they say and do.
- Offer chalk and black paper for pictures.
- Watch and note which children are able to observe carefully and see differences.
- Take some photos.

5. Taking it further

Try mixing:

- shaving foam and paint or soap flakes and water.
- baby lotion and powder paint to make face paint.
- flour, salt and water to make dough.
- instant whip pudding or instant custard.

Investigate reflections

Back to front and upside down. Investigate mirrors and reflections.

What you need:

- safety mirrors, as many sizes and shapes as you can manage
- plastic mirror sheet
- a tall mirror so children can see their whole body
- hand and handbag mirrors
- maybe a magnifying mirror

Key words

- reflect
- reflection
- light
- see
- in front
- behind
- up/down
- different
- back to front

Early Learning Goals

CLL – extend their vocabulary, exploring the meanings and sounds of new words

PSED – select and use activities and resources independently;

KUW – investigate objects and materials by using all of their senses as appropriate

– find out about, and identify some features of events they observe

CD – explore colour, texture, shape, form and space in two and three dimensions.

1. Explore

- Put mirrors in different places for the children to find and explore.
- Put some mirrors outside, and try fixing them to the ceiling, the floor or ground, to fences, bushes and trees.

2. Talk

- Talk about the things children can see in the different mirrors.
- Look carefully and describe what you can see.
- Look in mirrors on the ground and talk about the view.

3. Experiment and observe

- Use some small mirrors that can be carried around. Take them into different parts of your setting and garden.
- Put different things in front of the mirror and look at what is reflected. Try books, pictures, labels, photos.
- Talk about what you see, use the key words and encourage careful looking.
- Hang some small mirrors in bushes or from a fence. Look at what they reflect as they move.

4. Record

- Try drawing what you see in different mirrors.
- Make some notes of what different children say.
- Try painting on a big mirror and then taking a print on paper.
- Make a collection of pictures of different sorts of mirrors.

5. Taking it further

- Look at reflections in puddles and other water surfaces.
- Go on a mirror spotting walk, you'll be surprised how many you see!
- Look for shiny surfaces that reflect people or objects.
- Put some adhesive mirror tiles, mirror sheet or small mirrors in unexpected places such as very low down on the wall, under a tree.

Investigate magnets

The magic of magnets never fails to fascinate children and adults alike.

What you need:

- bar, horseshoe and pot magnets
- magnetic strip and sheet
- tins, nails, screws and clips
- sticks, pebbles, small plastic and wooden items
- clip or whiteboard
- camera

I will need

Key words

- stick
- attract
- metal
- wood
- different
- power
- strong
- pull

Early Learning Goals

CLL – extend their vocabulary, exploring the meanings and sounds of new words

PSED – be confident to try new activities
– select and use activities and resources independently

KUW – investigate objects and materials by using all of their senses as appropriate
– find out about, events they observe
– ask questions about why things happen and how things work

1. Explore

- Put the magnets and other objects on a table or other surface for the children to explore.
- Watch and listen to what they say and do.

2. Talk

- As they work, talk with them about what is happening. Join in and explore the magnets with them.
- Try picking up different objects.

3. Experiment and observe

- Now go on a magnet walk to see what you can find to pick up with the magnets.
- Talk as you work. Collect the objects in a basket or bag. Go outside as well as inside and see what you can find.
- When you get back, look at all the things the children have found and talk about why they think the magnet picks up some things and not others. What is the same about all the objects that 'stick' to the magnets?

4. Record

- Draw a circle on a whiteboard. Put all the things that the magnet attracts in the circle. Take a photo.
- Make a photo story or picture of your walk on a very long piece of paper. Draw the things you found along the walk.
- Make a picture list of things that are attracted to a magnet.

5. Taking it further

- Find some magnetic games to play.
- Collect some fridge magnets and play with them on an old tray.
- Use magnetic shapes and letters to make pictures or words.
- Use a magnetic strip or sheet to make your own games to play on an old fridge door.

Investigate glue

Look carefully at something the children often take for granted.

What you need:

I will need

- white glue
- spreaders
- brushes
- plastic sheeting (or bin liners cut open)
- tape
- food colouring

Key words

- stick
- sticky
- stuck
- clear
- transparent
- hard/soft
- dry
- wait
- watch
- feel
- peel
- bend

Early Learning Goals

CLL – extend their vocabulary, exploring the meanings and sounds of new words

PSED – be confident to try new activities
– select and use activities independently

KUW – investigate objects and materials by using all of their senses as appropriate
– look closely at similarities, differences, patterns and change.

1. Explore

- Tape some plastic sheet to a table.
- Pour some white glue onto the sheet and encourage the children to spread, squeeze and explore the glue.

2. Talk

- Talk with the children about how white glue behaves.
- Encourage the children to describe the feel, smell, sound of the glue as they work.
- Have a go yourself and see how it feels to you.

3. Experiment and observe

- Now observe the glue as it begins to dry. Look carefully at different places on the plastic where the glue is different thicknesses.
- Ask questions about what is happening.
- Keep watching as the glue dries. Keep talking about what happens.
- When the glue is completely dry, peel it off the plastic sheet. Talk about what happens and what the glue feels and looks like.

4. Record

- This investigation really needs photos to record the different stages of spreading and drying.
- Use the photos to make a book about the investigation and the things you do to follow it up.
- Make some notes of how individuals talk and observe.

5. Taking it further

- Add colour to some white glue and try the experiment again. What happens to the coloured glue?
- Investigate different sorts of glue – e.g. glue sticks, paste.
- Try putting torn tissue on spread out white glue on plastic sheeting. Peel off when it is dry to uncover a stained glass effect!

Investigate light and shadow

Provide some torches and investigate light and shadows.

What you need:

- torches
- a pop-up tent or homemade shelter or a blanket
- big sheets of white paper
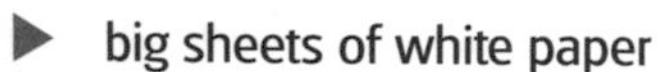
- objects to make shadows
- a spotlight or bedside light (or an overhead projector if you can borrow one!)

Key words

- light
- shadow
- shine
- switch
- dark
- point
- bulb
- on/off
- direction
- behind
- in front
- projector
- silhouette

Early Learning Goals

CLL – interact with others, negotiating plans and activities

PSED – be confident to try new activities, initiate ideas and speak in a familiar group

– work as part of a group or class

[illegible] – investigate objects by using all of their senses

– ask questions about why things happen and how things work

PD – use a range of small and large equipment

CD – respond in a variety of ways to what they see, hear, smell, touch and feel.

1. Explore

- Put a blanket or sheet over the tent to make it really dark inside.
- Offer the torches to the children, and let them experiment making shadows and light in the tent.

2. Talk

- Talk with the children about what happens in the dark. Discuss shadows and how they are made.
- Experiment again inside the tent, talking as you work with the children.

3. Experiment and observe

- Now let the children help you to experiment with shadows, holding up objects between the light and the tent wall.
- Hang some white paper or an old sheet in a dark place. Shine a strong light at the sheet and see how you can make shadows with your hand, your body, puppets or objects.
- Talk about what the light does, and how the features of an object disappear, leaving a silhouette.

4. Record

- Draw pictures of what you found out.
- Draw round shadows and silhouettes and compare them with the original object.
- Take some photos.
- Draw labelled diagrams of torches and how they work.

5. Taking it further

- Go outside on a sunny day and draw round shadows with playground chalk. Wait for an hour then draw round the shadow again – what has happened?
- Make a shadow puppet play. Draw characters on black paper. Cut them out and tape them to sticks. Use a strong spotlight or OHP to project yc puppets onto a sheet or white wall.

Investigate rolling

Use a plank or slide to investigate things that roll and things that slide.

What you need:

- guttering, a low slide or planks
- bricks to prop them up
- a collection of different balls (footballs, tennis, sponge, ping pong, koosh)
- clipboards, pens and chalk
- toy cars and trucks

I will need

Key words

- roll
- up/down
- fast
- slide
- roll
- wheels
- slope
- far/furthest
- stop/go
- best
- round
- flat
- push
- mark

Early Learning Goals

CLL – interact with others, negotiating plans and activities and taking turns in conversations
– extend their vocabulary, exploring the meanings and sounds of new words

) – select and use resources independently
– work as part of a group or class

KUW – investigate objects and materials by using all of their senses as appropriate
– build and construct with a wide range of objects, selecting appropriate resources
– ask questions about why things happen and how things work
– observe, find out and identify features in the natural world.

PD – use a range of small and large equipment.

1. Explore

- Put out the guttering and slopes. Add a basket of balls and some mark making equipment.
- If the children need some suggestions, start the investigation by rolling some balls down the slope.

2. Talk

- Talk with the children about what happens when the balls roll down the slide. Talk about balls that roll easily and balls that slide.
- Watch how far they go.

3. Experiment and observe

- Now make the game more systematic. Ask the children which ball they think will travel furthest when rolled down the slope.
- Test their guesses by rolling different balls down and marking where they stop with the chalk.
- Talk about what you find out. Test the balls more than once and record where they stop.

4. Record

- Use a clipboard or whiteboard to record your findings.
- Record what the children say, including new vocabulary.
- Take some photos of what happens. Try to catch the moving balls in your photo, and photograph the different sorts of balls you tested.
- Make a picture list of things that roll.

5. Taking it further

- Investigate toy cars and trucks. Roll them down slopes or along a flat surface. Don't forget to guess which will go furthest before you start your tests.
- Try the investigation again with a mixture of things that roll and things that slide (eg bricks, blocks, boxes and bean bags).

Investigate moving things

Shift, load and move object around your garden, and investigate moving things, levers, lifting and tipping.

What you need:

I will need

- trucks and barrows
- buckets, bags and baskets
- large stones or pebbles, bricks, boxes, small logs, small bags of sand, plastic crates and bottles.

N.B. These must be small enough for children to handle and lift, but heavy enough to need some exertion.

Key words

- lift
- carry
- full
- empty
- tip
- hold
- load
- push
- heavy
- big
- move
- pour
- strong

Early Learning Goals

CLL – interact with others, negotiating plans and activities and taking turns in conversations
– extend their vocabulary, exploring the meanings and sounds of new words

PSED – select and use resources independently
– work as part of a group or class

KUW – investigate objects and materials by using all of their senses as appropriate
– build and construct with a wide range of objects, selecting appropriate resources

PD – use a range of small and large equipment

1. Explore

- Put the stones or other materials in piles outside. Leave moving and carrying equipment nearby.
- Leave the children to play.
- Watch what happens.

2. Talk

- Join the children and talk about what they are doing.
- Use words like 'If ...'; 'I wonder ...'; 'Could you...?'
- Talk about what it feels like to be lifting and moving heavy things.

3. Experiment and observe

- Set up some simple challenges to move, fill, spread, pile, stack the objects.
- Test different ways of moving things and talk about which is easiest and which is most difficult.
- Look carefully at the different moving and carrying equipment. Which is easiest to use and why? Which is hard to use? Why? Which holds most? Which is easiest to load/unload?
- Record your findings.

4. Record

- Try a simple record by drawing the moving equipment and adding the number of things each will carry.
- Draw pictures with pencils or paint.
- Take some photos of the different vehicles and objects.
- Use a clipboard to make 'on the job' records.

5. Taking it further

- Make some slopes to walk or wheel up. Talk about the difference between moving things up slopes and on the flat.
- Make a builders' yard and add more objects to mix and move.
- Look for some pictures of levers and talk about how a lever works to help with lifting. Try some simple levers.

Investigate moving water

Make some waterways and explore how water moves. Start with just water, exploring how it behaves.

What you need:

- guttering and drainpipes
- duct tape, clips, pegs and string
- plastic bottles
- jugs, plastic jars and hose
- funnels, buckets and bowls
- bricks, blocks, A-frames etc. to prop the waterway up

Key words

pour	spill
fix	fill
stick	tip
up/down	hold
float	float
drip	run

Early Learning Goals

CLL – interact with others, negotiating plans and activities and taking turns in conversations;

PSED – continue to be interested, excited and motivated to learn
– select and use activities and resources independently

UW – investigate objects and materials by using all of their senses as appropriate
– find out about features of events they observe
– build and construct with a wide range of objects, selecting appropriate resources and adapting their work where necessary
– select tools and techniques.

1. Explore

- Leave all the items outside for the children to explore.
- Watch what happens. Help if asked, but try to just watch!
- Make suggestions if the children get stuck, but don't dominate the exploration.

2. Talk

- Begin to discuss what happens as the waterway develops.
- Talk about how to fix the guttering etc. together, what works, what doesn't and why.
- Watch and talk about the water and how it moves.

3. Experiment and observe

- Collect some small things that float (leaves, little boats, plastic containers, etc).
- Watch and talk as the children float things down the waterway. Draw their attention to how the water moves and how it moves things. Adjust the slope of the waterway to make the water go faster or slower.
- Watch what happens to the water as it comes out of the end of the waterway and spreads across the ground or falls into a bucket.

4. Record

- Take some photos as the waterway is built. Include photos of the experiments with sticking and making it waterproof.
- Record what children say as they experiment with the waterway.
- Offer long pieces of paper, pencils or fine felt pens for pictures.

5. Taking it further

- Look at water wheels and other water toys that make water move.
- Challenge the children to move a bucket of water from one side of the garden to the other using their waterway.
- Make a fountain, or buy a cheap fountain kit from a garden centre. Visit garden centres, parks, rivers and streams.

NB Always take extra adults when visiting rivers, ponds and streams to ensure safety.

Investigate floating and sinking

Make some of these water jars and watch how some things always return to the surface while others sink.

What you need:

I will need

- plastic or thick glass jars with closely fitting lids
- waterproof tape (duct tape)
- water, blue colouring
- a water tray or big bowl
- sand, gravel, stones and twigs, shells, cork, wood and polystyrene

Key words

- float
- sink
- bottom
- top
- in/on/under
- surface
- transparent
- wet
- soak
- pour
- waterproof
- leak
- turn/tip
- move

Early Learning Goals

CLL – interact with others, negotiating plans and activities and taking turns in conversations
– extend their vocabulary;

PSED – select and use activities and resources independently

KUW – investigate objects and materials by using all of their senses as appropriate
– find out about, and identify some features of objects and events they observe
– ask questions about why things happen

PD – use a range of small and large equipment.

1. Explore

- Place all the objects except the jars and tape near the water tray.
- Encourage free play and spend some time watching what the children do.

2. Talk

- Discuss what is happening, and begin to introduce the new vocabulary.
- Ask questions and listen to the answers about why some things float, some things sink, and some things stay under, but don't sink to the bottom.

3. Experiment and observe

- When the children have had plenty of time to explore, help them to put some water and blue colouring in the jars.
- Spoon a little sand into each jar – watch and discuss.
- Add some of the objects one at a time – watch what happens.
- When the children have put a range of different things in their jar, help them to seal the lid on their jar with tape.
- Now turn the jar over and see what happens. Turn or tip again, wait, watch and talk.

4. Record

- Take some photos of the free play and the jars.
- Record what the children say, including using the new vocabulary.
- Draw some jar outlines and record what happens when the jar is turned different ways.
- Make a picture list of things that float and things that sink.

5. Taking it further

- Collect or make some small boats and load them with plastic bricks, beads or other small items to see how much they can carry before they sink.
- Look at flotation aids such as arm bands, lifebelts, flotation suits.
- Find out about divers and submarines.

Investigate joining

This investigation is about joining materials. Focus their attention on the joins, not the finished model!

What you need:

- boxes, tubes, card, cartons, pots and other recycled materials. Have some big things and some smaller things
- several sorts of tape – duct, masking, selotape, insulating
- several sorts of glue

Key words

- stick
- cut
- tape
- glue
- fix
- end/side
- edge
- top/bottom
- together
- best
- dry
- hold
- help

Early Learning Goals

CLL – interact with others, negotiating plans and activities and taking turns in conversations

PSED – select and use resources independently

– work as part of a group

KUW – investigate objects and materials by using all of their senses as appropriate

– build and construct with a wide range of objects

– select tools and techniques they need to shape, assemble and join the materials they are using

CD – explore shape, form, space in three dimensions

1. Explore

- Leave all the things for the children to experiment with
- Listen and watch what happens and how the children work
- Encourage the children to experiment with all the different materials.

2. Talk

- Watch and discuss with the children how they are joining the different things.
- Help them to look carefully at which ways work best.

3. Experiment and observe

- Use a whiteboard or clipboard to record your findings as you discuss the best ways of fixing and joining things.
- Encourage the children to look at each others' work and see how other children have fixed things together.
- Talk about which glue and which tape sticks paper best, plastic best, card best. Ask which tapes and glues are easiest to use.
- Try the joins again when the constructions are painted.

4. Record

- Record some children as they talk about their experiments, or take some notes of what they say.
- Give the children paper and fine felt pens to draw what they have found out.
- Take some photos of joins.

5. Taking it further

- Try leaving the joined things in different places (such as outside, in a fridge), and see what happens to the glue.
- Look at the ways different things are joined – boxes, bags, books, clothes, shoes, toys etc.
- Look at joining toys, such as velcro toys and magnets.

Investigate lifting things

Lifting things is a scientific activity. Children need practical experiences to learn about pulleys and levers.

What you need:

- planks
- rope and string, clips and hooks etc
- baskets and boxes
- teddies and soft toys
- plastic and wooden bricks
- A frames, climbing frames, steps, etc.

Key words

- lift
- pull
- heavy
- light
- up/down
- high
- load
- basket
- plank
- frame
- over/under
- through

Early Learning Goals

CLL – interact with others, negotiating plans and activities and taking turns in conversations

– extend their vocabulary, exploring the meanings and sounds of new words

SED – select and use resources independently

– work as part of a group or class

KUW – investigate objects and materials by using all of their senses as appropriate

– build and construct with a wide range of objects, selecting appropriate resources

PD – use a range of small and large equipment

1. Explore

- Leave all the things outside for the children to experiment with.
- Watch what happens. Listen to what they say and watch how they work together to use the things you give them.

2. Talk

- Talk with them about what they are doing. Try not to interfere, and only help if they ask you to!
- Use 'I wonder...' and 'What do you think ...' and 'How could you?'

3. Experiment and observe

- When they have had some time for free experimentation, give the children some challenges.
- Challenge them to lift a teddy in a basket to the top of the climbing frame, or move a box of bricks up a step.
- Talk as they try the challenges. Make suggestions, but don't take over!
- Take some action photos as they work, and note how they collaborate on tasks.

4. Record

- Use the photos as discussion starters. Make them into a book and write what the children say under each photo.
- Draw the constructions and solutions.
- Collect construction pictures from catalogues and magazines for an ideas scrapbook.

5. Taking it further

- Watch a DVD or visit a building site to look at lifting and moving loads with cranes and diggers.
- Put strings, cranes and pulleys in the sand or construction trays.
- Use small world diggers, tippers, cranes and lorries to move sand, stones and gravel around.

NB Make sure children don't strain muscles by carrying things that are too heavy.

Investigate things that turn

Collect and explore all sorts of things that turn and twist.

What you need:

- screw top jars
- doorknobs, keys and locks
- clockwork toys, clocks with keys
- knobs, buttons and dials
- screw toys
- nuts, bolts, screws, screwdrivers

You could fix some of these things to a board so they can be turned and twisted

Key words

- turn
- twist
- on/off
- louder
- softer
- up/down
- switch
- fix
- control
- screw
- wind

Early Learning Goals

CLL – extend their vocabulary, exploring the meanings and sounds of new words

KUW – investigate objects and materials by using all of their senses as appropriate
– look closely at differences and change
– ask questions about why things happen and how things work
– observe, find out and identify features in the natural world.

PD – use a range of small equipment
– handle objects safely and with increasing control.

1. Explore

- Leave all the things for the children to experiment with freely.
- Watch what happens. Look at how the children use their fingers and wrists as they turn and twist.

2. Talk

- Join the children as they work. Ask open questions about what they have found out.
- Talk about where the objects came from, how they work and what they do.

3. Experiment and observe

- Go on a 'Twisting, turning things walk.' Look for anything that twists or turns.
- Stop whenever you find a turning object and talk about it.
- Take some photos as you go.
- Try taking a lock apart to investigate how it works (if you ring a local locksmith, he may be able to give you some old locks, and an ironmonger or a DIY place may donate some display doorknobs, old nuts and bolts, etc.)

4. Record

- Record what the children say as they work or try things out.
- Offer fine felt pens or pencils to record in pictures.
- Make a picture list of things that twist and turn. Use junk mail and DIY catalogues.
- Take some photos on your walk. Display them with some turning things.

5. Taking it further

- Make a control panel for a rocket, a car or an aeroplane. Fix knobs and dials to a piece of wood or even a sturdy cardboard box. Use for role-play.
- Try some simple turning tools and toys - a football rattle, a rotary whisk, a toy windmill, a water wheel, a musical box or Etch-a-Sketch.

Investigate bending things

This investigation is about materials that twist, bend and flex. The activity focuses on construction.

What you need:

- strips of paper, plastic, card (different colours, thicknesses etc.
- cardboard tubes
- pieces of card cut from the sides of cardboard boxes to use as bases for the constructions
- glue, staplers, tape, clips and scissors

Key words

- bend
- twist
- fold
- round
- stick
- fix
- spring
- bounce
- glue
- cut
- best

Early Learning Goals

CLL – interact with others, negotiating plans and activities and taking turns in conversations;

PSED – select and use resources independently
– work as part of a group

KUW – investigate objects and materials by using all of their senses as appropriate
– build and construct with a wide range of objects
– select tools and techniques they need to shape, assemble and join the materials they are using

1. Explore

- Talk to the children about the investigation and introduce the materials.
- Sit with the children and watch what happens as they work.

2. Talk

- Encourage the children to twist and turn the strips of paper, and bend the tubes. If they find it difficult, model how to do it, but don't dominate.
- Listen to what they say as they work.

3. Experiment and observe

- Now try making a group construction. Put a big piece of cardboard box on the table and start to work with the children, bending and sticking the strips and tubes to make them stand up in hoops and loops, twisting some and leaving others flat. The construction may become very complex as you all work.
- Stop every so often and look at your construction. Talk together about what you have done and how it works.

4. Record

- Take photos as you work together. Use close up pictures to catch children's hands and wrists in action, and to capture details.
- Record what the children say as they work.
- Draw pictures of your constructions with pencil or paint.

5. Taking it further

- Get hold of an 'igloo' tent or 'pop up'. Look at how it stands up.
- Look at pictures of structures – bridges, tunnels, domes, etc.
- Make some willow structures outside by bending willow branches and pushing the ends in the ground. Cover with fabric or creeping plants. You could try living willow, which grows its own leaves!

NB You can get living willow from some garden centres or specialist suppliers (see resources section).

Investigate boxes

Collect plenty of sorts and sizes of boxes before you start this investigation into how they are made.

What you need:

- lots of different boxes
- scissors
- chalk
- glue and masking tape
- flat card
- pencils and pens

Key words

- box
- carton
- bottom
- top
- side
- shape
- square
- rectangle
- inside
- flat
- stick
- outline
- draw

Early Learning Goals

CLL – interact with others, negotiating plans and activities and taking turns in conversations

PSED – be confident to try new activities, initiate ideas and speak in a familiar group

KUW – investigate objects and materials by using all of their senses as appropriate

– ask questions about how things work

– build and construct with a wide range of objects

– select tools and techniques they need to shape, assemble and join the materials they are using

PD – use a range of small and large equipment.

1. Explore

- Put some boxes on a table or the floor (inside or outside).
- Introduce the activity to the children. Tell them it is about finding out how boxes are made.

2. Talk

- Talk with the children as they explore the boxes, looking carefully inside and out, taking them apart and flattening them.
- Encourage them to look carefully at the joins.
- Help with any staples or clips.

3. Experiment and observe

- Now experiment with the flattened boxes.
- Look at the shapes, naming each one. Talk about the sides and the bottoms of the boxes. Feel the folds, corners and joins.
- Look at how different boxes are put together and fixed.
- Try putting the flattened shapes back together again with tape or glue.
- Draw round the flattened boxes and make some more the same.

4. Record

- Record what the children say, including their comments on how boxes are made.
- Draw pictures of unfolded boxes with pencil or fine felt pen.
- Take some 'before' and 'after' photos of some of the boxes.
- Make a display of all the different boxes you have investigated.

5. Taking it further

- Look at nesting boxes or sets of different sizes.
- Experiment with fitting a range of boxes of different sizes inside each other.
- Use empty boxes to make buildings and constructions.
- Use some very big boxes to make vehicles and houses.

Investigate wheels

Look at wheels everywhere - how they move, how they work, how they fit, what they are made of.

What you need:

- outdoor wheeled toys – trikes, bikes, prams and scooters
- toy cars and other small wheeled vehicles
- construction kits with wheels and cogs
- plasticene, paint, dough and paper

Key words

- wheel
- round
- turn
- tyre
- axle
- bolt
- pin
- puncture
- tracks

Early Learning Goals

CLL – interact with others, negotiating plans and activities and taking turns in conversations

PSED – be confident to try new activities
– select and use activities and resources independently

W – investigate objects and materials by using all of their senses as appropriate
– ask questions about why things happen and how things work

PD – use a range of small and large equipment
– handle tools, safely and with control

1. Explore

- Start indoors or out of doors with indoor or outdoor wheels.
- Explain the activity to the children, and encourage them to look carefully at the wheels of the toys and how they work.

2. Talk

- Join the children as they investigate the wheels. Listen to what they say and what they have found out.
- Ask open questions and suggest other things to look at.

3. Experiment and observe

- Now compare the wheels you have been looking at with others (go outdoors or come indoors).
- Look for similarities and differences in the wheels – their size, shape, the way they are fixed.
- Now look at the construction sets and discuss the way the wheels work there.
- Try making tracks in clay, paint or dough with all the different wheels. Look for similarities and differences in the tracks.

4. Record

- Make patterns or prints with wheels.
- Draw pictures with pencil or fine felt pen.
- Collect pictures for a scrapbook of things with wheels.
- Go for a walk and tally the number of vehicles and other things with wheels that you see.

5. Taking it further

- Visit a garage or tyre fitting place.
- Bring an adult bike into your setting and let the children explore and draw it.
- Add thick card wheels to the technology area and see what the children build.

Investigate leaves

Look carefully at leaves and investigate similarities and differences in this investigation for all seasons.

What you need:

- a selection of leaves from bushes, trees, plants, weeds and grasses
- magnifying glasses
- crayons, pencils and pens
- thin paper
- clipboards and pegs

Key words

- shape
- colour
- sections
- edges
- shiny
- smooth
- bumpy
- stalk
- veins
- dark/light
- soft
- hard
- prickly
- spiky

Early Learning Goals

CLL – extend their vocabulary, exploring the meanings and sounds of new words

SED – continue to be interested, excited and motivated to learn
– work as part of a group or class

– investigate objects by using all of their senses
– find out about and identify some features of living things
– look closely at similarities, differences and change
– observe the natural world

PD – use a range of small and large equipment

1. Explore

- Go into the garden or park and collect some leaves. Give each child a basket or bag for their own collection.
- Don't forget to collect grass, weeds and tree leaves.
- Bring the leaves back and spread them on a table or board, indoors or out, so you can all look at them.

2. Talk

- Join the children and talk about what they have found. Begin to explore the different leaves using the new words

3. Experiment and observe

- Give the children some magnifying glasses to examine the leaves.
- Encourage them to tell you what they can see and feel.
- Try sorting the leaves into different types, sizes, textures and colours.
- Talk about the different parts of the leaves – veins, stalks, edges and shapes.
- Look carefully at the different surfaces of the leaves. Talk about the differences between the back and front.

4. Record

- Take some photos of the plants, bushes and trees, and some close up photos of some of the leaves.
- The children could draw pictures of what they see through the magnifying glasses.
- Help the children to record some of their findings in pictures or words.

5. Taking it further

- Make some leaf prints by painting the leaves with thick paint and pressing them down on paper, or put the leaves under thin paper and rubbing on the paper with the side of a crayon.
- Talk about what leaves are for and how they help plants to collect energy from sunlight.

Investigate seeds

This investigation can be done at any time of year, but spring and summer are best, when you can grow the seeds.

What you need:

I will need

- seeds of different types – tree seeds, nuts, flower and vegetable seeds, beans, peas, bulbs, pips and stones from fruit and vegetables
- pots, seed trays, compost and water
- jars and blotting paper or cotton wool

Key words

- seed
- nut
- bean/pea
- pod
- grow
- roots
- leaves
- stalk
- flower
- food
- water
- sunlight
- plant
- wait

Early Learning Goals

CLL – extend their vocabulary, exploring the meanings and sounds of new words

PSED – continue to be interested, excited and motivated to learn
– work as part of a group or class

JW – investigate objects by using all of their senses
– find out about, and identify some features of living things
– look closely at similarities, differences, change
– observe the natural world

PD – use a range of small and large equipment

1. Explore

- Put the seeds on a table or tray for the children to explore. Leave some in their packets or leave them in cut fruit or pods.
- Watch how the children explore, sort, examine and handle the seeds.

2. Talk

- Join the children as they explore and talk with them about the different sorts of seeds. Offer some saucers or trays for sorting.
- Introduce some new words.

3. Experiment and observe

- Add some magnifying glasses for the children to use.
- Continue to look at the differences and similarities in the different sorts of seeds.
- Talk about where they come from and what they do.
- Provide some pots and compost or some cotton wool and jars so they can plant some seeds.
- Talk about what seeds need to grow (water, warmth, light, soil).

N.B. Some of the most rewarding seeds to plant are cress, beans and peas in jars, nasturtiums, grass seed, sunflowers, mung beans.

4. Record

- Encourage children to draw or paint pictures of the different sorts of seeds and where they come from.
- Record what the children say, including use of the new vocabulary.
- Photograph the planting of seeds and make a 'Growing' photo book.
- Find pictures of flowers and seeds in magazines and catalogues.

5. Taking it further

- Watch the seeds as they grow. It's quite alright to dig them up gently to see what they are doing, as long as you don't squash them!
- Make a chart with the days on and draw what happens to the seeds as they begin to grow. Plant them out in a growbag or pot.
- Plant 20 peas and dig up one each day to see what has happened.

Investigate minibeasts

Use your garden, the local field or park for this investigation into the lives of small creatures.

What you need:

- small, clear jars with lids
- bug collectors, Pooters or Bug Jars (see end of book for suppliers)
- magnifying glasses
- camera
- bug spotting book
- clipboards or whiteboards

Key words

- spider
- beetle
- ladybird
- worm
- snail
- slug
- ant
- greenfly
- wasp
- bee
- woodlouse
- creature
- legs
- catch
- under
- tickle
- careful
- hole
- nest
- scared
- touch
- gentle
- run
- jump

Early Learning Goals

CLL – extend their vocabulary, exploring the meanings and sounds of new words

PSED – continue to be interested, excited and motivated to learn

KUW – find out about features of living things

– observe, find out and identify features in the place they live and the natural world

PD – use a range of small and large equipment

CD – respond in a variety of ways to what they see, hear, smell, touch and feel

1. Explore

- ▶ Talk to the children before you collect the minibeasts. Emphasise care and kindness and the need to be still and quiet.
- ▶ Take the bug collectors out and see what you can find!
- ▶ Bring the minibeasts back to your setting and look at them carefully.

2. Talk

- ▶ As you look, talk with the children about the different creatures, their names and features.

3. Experiment and observe

- ▶ Introduce some simple reference books and magnifying glasses. Encourage the children to look carefully at the creatures they have found, counting legs, wings and eyes, naming colours and movements. Find the creatures in the books.
- ▶ Talk about where different minibeasts live, what they eat, how they move.
- ▶ Don't forget to let them go on the same day you catch them.

4. Record

- ▶ Use fine felt pens or pencils to draw the minibeasts.
- ▶ Record what the children say, and their likes and dislikes.
- ▶ Take some photos of the minibeast hunt and the observation.
- ▶ Make picture lists or sets of minibeasts with six legs, eight legs, wings, eyes on stalks etc.

5. Taking it further

- ▶ Make a terrarium in a plastic aquarium or big plastic box. Line it with soil and a bit of real turf if possible. Add some leaves, sticks, stones and gravel, a plastic lid full of water and some of the minibeasts you have collected. Don't forget to put a cover on or the minibeasts will escape! Only keep the minibeasts for a day or two, then release them and collect some new ones.

Investigate butterflies

This investigation needs a bit of preparation and the observations take place over time.

What you need:

- a butterfly garden (see end of book for details)
- magnifying glasses
- 'The Very Hungry Caterpillar' book
- life cycle books
- fine felt pens and clipboards
- a calendar

Key words

- eggs
- caterpillar
- cocoon
- butterfly
- hatch
- change
- legs
- wings
- fly
- crawl
- wriggle
- eat
- drink
- wait
- release
- garden
- leaves
- sleep
- inside
- free
- touch
- gentle
- care
- tickle

Early Learning Goals

CLL – extend their vocabulary, exploring the meanings and sounds of new words

PSED – continue to be interested, excited and motivated to learn

KUW – find out about change in living things

– observe, find out and identify features in the place they live and the natural world

PD – use a range of small and large equipment

CD – respond in a variety of ways to what they see, hear, smell, touch and feel

1. Explore

- Read The Very Hungry Caterpillar and talk about the story.
- Explore other stories and fact books about caterpillars and butterflies.
- Sing the Caterpillar Song (see page 66).

2. Talk

- Talk about life cycles.
- Make and play some sequencing games of life cycles.
- Go outside or to the park and look for caterpillars and other minibeasts.
- Order your butterfly box.

3. Experiment and observe

- When your butterfly box arrives, open it with the children and talk about what will happen.
- Use a calendar to discuss the length of time the caterpillars will need to turn into butterflies.
- Make a calendar to record your observations and how the caterpillars develop.
- Look at the caterpillars every day and talk about how they grow and change.
- When the butterflies hatch, let them go in your garden.

4. Record

- Take some photos. A series of photos of the development of the caterpillars would make a good sequencing game.
- Record the growth of the caterpillars, cocoons and butterflies in pictures on a day by day chart. Write what the children say under each picture.

5. Taking it further

- Go into the garden and watch out for butterflies in the summer.
- Make a picture list of creatures that can fly, or creatures that change during their life cycle (eg moths, beetles, ladybirds).
- Try making a ladybird box or a wormery.
- Make your own book version of 'The Very Hungry Caterpillar'.

Investigate under the ground

Find a place to dig, shovel and move the earth so you can explore what is under the ground.

What you need:

- spades, forks, trowels and rakes (child-sized if possible)
- plastic trays, buckets, bowls, boxes, jars and sieves etc.
- bug boxes and magnifying glasses
- a camera

Key words

- dig
- turn
- lift
- look
- uncover
- hole
- carry
- grass
- stones
- worm
- insect
- beetle
- rocks
- alive

Early Learning Goals

CLL – interact with others, negotiating plans and activities and taking turns in conversations

PSED – select and use activities and resources
– work as part of a group

KUW – build and construct with a wide range of objects
– observe, find out and identify features in the place they live and the natural world

CD – respond in a variety of ways to what they see, hear, smell, touch and feel

1. Explore

- Introduce the activity to the children. Use the correct names for the tools and model how to use them safely.
- Show them the place where they can dig and make sure they understand any rules.

2. Talk

- As the children work, talk about the sorts of things they might find.
- Watch to see how they manage the tools.
- Encourage them to stop when they find something.

3. Experiment and observe

- Now talk about what they have done so far. Stand back and look at the holes and dips they have dug. Look at some of the things they have found.
- Discuss whether each thing is alive or not.
- Talk about how to look after the things they find, particularly anything living.
- If you want to keep worms (most children will!) put them in a bowl with some damp soil, leaves and turf.

4. Record

- Take some photos of the activity and the finds.
- Record what the children say, including how they use any new vocabulary.
- Make an under the ground museum or display and label all the items you have found. Make tickets and a catalogue.

5. Taking it further

- Sort the objects they have dug up into living and not living.
- Talk about holes and digging, visit a building site, look out for workmen mending the road.
- Get some simple books about underground and the things that happen there – e.g. underground trains, archaeology, tunnels, caves, dinosaurs, bats, basements and cellars.

Investigate the weather

This is another extended investigation. You could make it as short as a week or as long as a term.

What you need:

To make your weather station:

- card, scissors and a pen
- a garden cane
- an empty plastic bottle
- a broom handle
- a bucket of sand
- playground chalk and ribbons

Key words

- weather
- rainy
- windy
- sunny
- snowy
- foggy
- cloudy
- misty
- shadow
- rain gauge
- windmill
- puddle
- raindrop
- rainbow

Early Learning Goals

CLL – interact with others, negotiating plans and activities and taking turns in conversations

PSED – continue to be interested, excited and motivated to learn

KUW – find out about, and identify some features of events they observe

– look closely at similarities, differences, patterns and change

– ask questions about why things happen

– find out about their environment, and talk about those features they like and dislike

1. Explore

- Talk about the weather every day and help the children to make a weather chart on a big whiteboard or card.
- Informally ask individual children about the weather as they come in the morning

2. Talk

- At group time and every time you go out, talk about the weather, using weather words.
- Record some weather news from the TV or radio. Make a TV from a box and make a real weather report.

3. Experiment and observe

- Now help the children to make a weather station for your garden.
- Make a windmill and hang some ribbons up for the wind.
- Cut the top off a plastic bottle, turn it upside down and put it back to make a rain gauge.
- Make a sundial from a broom handle in a bucket of sand and watch the shadow move.
- Use a clip board to make daily weather reports, and talk about what you find.

4. Record

- Take some photos on different days and see how different your setting looks in different weathers.
- Make a weather chart for a week or even for a whole month.
- Draw pictures of different sorts of weather.
- Cut out weather pictures from magazines to illustrate your findings.

5. Taking it further

- Watch raindrops running down windows and into puddles.
- Make kites and card windmills to fly in the wind.
- Put an unbreakable mirror on the ground so the children can see the sky and the clouds.
- Get some travel brochures and look at weather in other lands.

Investigate holes

There are holes everywhere! You can investigate small ones in everyday objects or big ones in the ground.

What you need:

I will need

- things with holes – stones, shells, corks, buttons, strainers, colanders, ring pulls, beads, rings, bracelets, pasta tubes etc.
- hole punches, strings and laces
- threading toys
- spades, shovels and buckets

Key words

- through
- thread
- hole
- inside
- dig
- in/out
- under
- dark
- down
- den
- cave

Early Learning Goals

PSED – continue to be interested, excited and motivated to learn

KUW – investigate objects by using all of their senses
– look closely at similarities and differences

PD – use a range of small and large equipment
– handle tools, objects, safely and with increasing control

CD – explore colour, texture, shape, form and space in two and three dimensions.

1. Explore

- Put all the objects with holes in a box or basket. Add some strings and laces, and offer as a free choice activity to the children.
- Sit with the children and watch what happens as the children explore the holes in different objects.

2. Talk

- Talk with the children about the different sorts of holes, how they got there, and what they are for.
- Talk about other holes.

3. Experiment and observe

- Go on a 'Hole Spotting Walk'. Take a camera and snap some of the holes. Look at manhole covers, keyholes, letter boxes, drainpipes, holes in walls, gutters, as well as holes made by animals and insects.
- Talk about what the holes are for and who or what made them.
- Make holes in mud with sticks.
- Experiment with making holes in clay, sand or dough, using cutters and sticks.

4. Record

- Record your 'Hole Spotting Walk' on a very long piece of paper, drawing the walk and all the holes you found.
- Draw pictures of things with holes using pencil or fine felt pen.
- Make a picture list of things with holes.
- Make strings of things with holes and hang them up.

5. Taking it further

- Investigate creatures that live in holes – rabbits, foxes, badgers, mice, bumble bees, toads, woodpeckers and kingfishers etc.
- Make a display of things with holes, ask parents to contribute ideas.
- Investigate pouring liquids through things with holes – tubes, funnels, colanders, sieves and strainers.

Investigate fabrics

Investigate colour, texture and other characteristics of fabrics.

What you need:

- a range of fabrics of different textures and lengths - e.g. fur fabric, silk, wool, cotton, voile, leather, crepe, denim, velvet, fleece, knits, towel, lycra, lurex, plastic and bubble wrap
- clothes pegs
- clipboards

I will need

Key words

- soft
- smooth
- shiny
- bumpy
- slippery
- hairy
- special
- wedding
- party
- birthday

Early Learning Goals

CLL – extend their vocabulary

PSED – be confident to try new activities, initiate ideas and speak in a familiar group
– select and use resources independently

KUW – investigate materials by using all their senses
– look closely at similarities, differences

PD – handle tools, objects with increasing control

CD – explore colour, texture, in 2 and 3 dimensions
– respond in a variety of ways to what they see, hear, smell, touch and feel

1. Explore

- Put the fabrics in a basket or box and leave for the children to explore.
- Watch what happens.
- Listen to what the children say and do as they feel, use and wear the fabrics.

2. Talk

- Talk with the children about the different textures and colours of the fabrics.
- Talk about who might use or wear them, and which ones are for special purposes.
- Look at pictures of clothes.

3. Experiment and observe

- Now work with the children to investigate the properties of the fabrics.
- Talk about which fabrics are good for keeping you warm, which ones would keep you dry, which ones stretch, or fray, or unravel.
- Test some fabrics for water resistance, stretch, warmth, strength e.g. pour drops of water on the fabrics. See which ones soak up the water and which ones repel it.

4. Record

- Sort and label the fabrics into 'shiny/dull', 'smooth/rough', 'ordinary/special' or into colours.
- Note what the children say, including new or unusual vocabulary.
- Collect pictures of clothing and other fabrics.
- Ask parents to donate left over fabrics for your collection.

5. Taking it further

- Make a tactile patchwork picture with shapes of different fabrics.
- Do some weaving to make your own fabric.
- Try un-making fabrics. Fringe them, unravel them, fray them.
- Visit a fabric shop or street market to look at fabrics.

Investigate bodies

Help children look carefully at their own bodies, and compare and contrast them with others.

What you need:

- mirrors
- big sheets of paper (rolls of wallpaper or lining paper work well)

- chalk, paint, scissors
- camera
- magnifying glasses

I will need

Key words

- skin
- colour
- eyes
- eyebrows
- eyelashes
- cheeks
- freckles
- chin
- teeth
- lips
- hands
- feet
- nails
- hair
- same
- different
- curly
- straight
- long
- short
- tall

Early Learning Goals

CLL – extend their vocabulary, exploring the meanings and sounds of new words

PSED – work as part of a group or class, taking turns and sharing fairly, understanding that there need to be agreed values

KUW – find out about, and identify some features of living things

– look closely at similarities, differences

CD – respond in a variety of ways to what they see, hear, smell, touch and feel

1. Explore

- Put out the mirrors and some paper and pencils.
- Explain the activity to the children. Say you are going to look at yourselves and each other.
- Take a photo of each child, so they can refer to it.

2. Talk

- Sit with the children and help them to look carefully at their own faces and bodies. Listen as they name features of their faces, hair, eyes and teeth etc.

3. Experiment and observe

- Now help the children to draw round each other on big sheets of paper and paint their own features. Stay close, use the mirror, and talk them through getting the features accurate.
- Help to cut them out and pin them up where you can compare features such as height and hair colour etc.
- Talk about similarities and differences between children. Do this in pairs or small groups.

4. Record

- Make a photo album with all the photos and words that the children dictate (or write if they can).
- Put the mirrors where the children can use them to draw pictures of themselves.
- Collect pictures of faces and people from magazines and catalogues.

5. Taking it further

- Use the information from the cut out children to make block graphs and pictograms.
- Use people jigsaws, posters and books for discussion starters.
- Sing body songs and rhymes such as 'Put your Finger in the Air', 'Heads, Shoulders, Knees and Toes'.

Investigate sounds

Start with sound makers and move on to simple instruments in an investigation of sounds all around.

Remember that all sound work needs managing, especially free exploration. Make sure you have planned for the inevitable noise!

What you need:

Group 1 – sound makers:

- boxes, pans, tins with rice or stones in, chains, keys, spoons, rattles, sticks and other simple sound makers

Group 2 – simple instruments:

- tambourines, drums, triangles, chime bars, bells, maracas, rain sticks and other simple instruments

Key words

- shake
- rattle
- bang
- tip
- slap
- note
- music
- tune
- click
- crash
- tinkle
- shiver
- ting
- beater
- stick
- high/low
- loud/soft
- long/short
- listen
- turn
- wait

Early Learning Goals

CLL – interact with others, negotiating plans and activities and taking turns in conversations
– extend their vocabulary

PSED – work as part of a group or class

KUW – investigate objects and materials by using all of their senses as appropriate

CD – respond to what they hear
– recognise and explore how sounds can be changed, sing simple songs

1. Explore

- Start by taking a 'listening walk' around your setting or outside. Stop frequently to listen and discuss what you hear.
- Talk about the walk, then put some Group 1 sound makers in a basket and offer to the children.
- Watch how they experiment with the different sounds.

2. Talk

- Join the group and play along side, talking about the sounds.
- Encourage careful listening.

3. Experiment and observe

- Add some of the Group 2 simple instruments to the activity.
- Play them again.
- Encourage the children to listen carefully to the sound each one makes. Use descriptions and comparisons (see key word list).
- Make up some simple sound patterns using all the different sound makers. Encourage turn taking and careful listening.
- Use your sounds to accompany familiar songs.

4. Record

- Make a recording of your 'listening walk'. Use it as a reminder of the walk. Take some photos of the sources of sounds.
- Record what the children say as they experiment with the sounds.
- Draw pictures and patterns of the sounds you make.
- Photograph the sound makers and use them as picture clues.

5. Taking it further

- Make some simple instruments (shakers, drums, rattles).
- Listen to unusual sound makers – such as a hooter, a toy horn, a bike bell, a kitchen timer, an electric toothbrush.
- Listen for sounds around your setting – the computer bleeping, the phone ringing, a clock ticking, the tap or toilet flushing.

Investigate textures

Go outside your setting, to a garden, park or field, and start feeling the world around you.

What you need:

- magnifying glasses
- thin paper
- camera
- carrier bags
- clipboards
- pens or pencils

Key words

- rough
- smooth
- bumpy
- hard
- soft
- tickly
- crunchy
- cold
- touch
- warm
- pattern
- texture

Early Learning Goals

CLL – interact with others, negotiating plans and activities and taking turns in conversations
– extend their vocabulary, exploring the meanings and sounds of new words

KUW – investigate objects and materials by using all of their senses as appropriate
– find out about, and identify some features of living things, objects and events they observe
– look closely at similarities, differences, patterns and change
– observe, find out and identify features in the place they live and the natural world.

1. Explore

- Invite the children to join you on a 'feeling walk.'
- Go outside and feel all the different textures. Most children find it easier to close their eyes when they feel things.

2. Talk

- Listen to what they say about the feel of the objects they find. Encourage them to feel carefully and describe what they can feel.
- Encourage use of new vocabulary.

3. Experiment and observe

- Collect some of the textured things you find and bring them back to your setting.
- Take some photos of things that are too big, or otherwise too difficult to bring back (e.g. a cold lamp post, bumpy tree bark, growing plants).
- Talk about the different textures and patterns, sort the objects into different categories.
- Discuss how different objects become textured in different ways.

4. Record

- Take some photos.
- Record what the children say, including new vocabulary.
- Sort objects and label them with texture names.
- Make a display of textures.
- Make some texture rubbings with thin paper and wax crayons.

5. Taking it further

- Make a 'feely box' and try to guess what is inside just by feeling.
- Try blindfolds – a soft scarf, headband, or toy mask. Take turns to feel another person's face or an object and guess who or what it is (N.B. some children are very frightened of being blindfolded!).
- Make textured dominoes or a texture matching game.

Investigate fruit and vegetables

Get different fruit and vegetables to try at snack time, and explore them as you prepare and eat them.

What you need:

I will need

- different fruits – e.g. apple, pear, mango, banana, grape, kiwi fruit, passion fruit, strawberry, star fruit, blackberry and avocado
- different vegetables – e.g. celery, tomatoes, cucumber, carrots, lettuce, peas, onions etc.
- blunt knives and magnifying glasses

Key words

- colour
- crisp
- crunchy
- sweet
- sour
- juicy
- skin
- shape
- smell
- taste
- pips
- seeds
- inside
- outside

Early Learning Goals

CLL – extend their vocabulary

PSED – be confident to try new activities, initiate ideas and speak in a familiar group
– work as part of a group or class

KUW – investigate by using all of their senses
– find out about, and identify some features of things they observe
– look closely at similarities, differences
– find out about the natural world

CD – respond in a variety of ways to what they see, hear, smell, touch, taste and feel

1. Explore and Talk

- Buy a range of fruit and vegetables that can be eaten raw.
- Every day, look carefully at a different fruit or vegetable. Talk about the skin, stalk, colour, texture.
- Look, feel and smell each before you cut or peel it.
- Now look carefully at the inside of the fruit/vegetable. Notice any seeds or pips and where they grow.
- Try to keep one of each sort of fruit and vegetable whole so you can use them all together later in the week.

2. Experiment and observe

- Later in the week, put all the fruit and vegetables out together as an activity.
- Join the children while they look, feel, explore, cut up and taste the whole range of fruit and vegetables.
- Try sorting into fruit and vegetables, ones with stones and ones with pips, soft ones and crunchy ones, different colours.
- Keep some of the pips and seeds to grow (avocado, apple, orange work well).

3. Record

- Draw or paint pictures of the whole and cut fruits and vegetables.
- Collect pictures of fruit and vegetables from magazines or junk mail.
- Make a picture list of fruit and one of vegetables.
- Photograph each snack time and the final investigation – make a book of your photos.

4. Taking it further

- Use some fruit and vegetables to make prints. Try onions, apples, carrots, cucumber. Stick the pieces on forks to make printing easier.
- Visit a local market or greengrocer.
- Explore some more unusual fruits.
- Investigate the Healthy Schools Initiative. See end of this book.

The Caterpillar Song

Sing it to the tune of 'She'll be Coming Round the Mountain'.

There's a tiny caterpillar on a leaf (wiggle wiggle)
There's a tiny caterpillar on a leaf (wiggle wiggle)
There's a tiny caterpillar, a tiny caterpillar, a tiny caterpillar on a leaf (wiggle wiggle, chomp, chomp, spin, spin, flap, flap).

Arabella Miller

Little Arabella Miller
Found a furry caterpillar
First she put it on her mother
Then upon her baby brother
He said, 'Arabella Miller,
Take away that caterpillar!'

Resources and Addresses

Bug Boxes, etc.

Insect Lore, PO Box 1420, Kiln Farm, Milton Keynes, MK19 6ZH.

www.insectlore-europe.com – bug viewers, butterfly and other insect boxes, books and other minibeast things.

Gardening tools for children

BRIO – gardening tools and children's gardening gloves (they also do waterways)

GONE GARDENING – **www.gonegardening.com** – for children's tools

GARDENLINKS UK – **www.gardenlinks.co.uk** – children's gardening and tools

Healthy Schools Initiative

The Healthy Schools Initiative - guidance on a range of health issues including healthy eating and Fruit for Schools.

– **www.DfES.gov.uk** gives general information.

www.healthyschools.gov.uk gives more information on Healthy Schools for parents, teachers and children, including a range of playground activities.

Contact your local education department to find out who is co-ordinating the initiative in your area.

Light boxes

ALC Associates make and supply light boxes and light tables, modelled on those used in Reggio Schools.

Contact them at **www.alcassociates.co.uk**

Magnets etc.

ASCO – magnets, magnifiers, height charts and measures, 'A' frames for climbing and making slopes, simple percussion instruments, mirrors. – **www.ascoeducational.co.uk**

Willow

Suppliers of living willow: PH Coates & Son, Meare Green Court, Stoke St Gregory, Taunton, Somerset, TA3 6HY. – **www.englishwillowbaskets.co.uk**

through the EYFS

Great for the Early Years Foundation Stage!

Ideal to support progression and extend learning.

If you have found this book useful you might also like ...

LB Making Poetry

ISBN 978-1-4081-1250-2

LB Christmas

ISBN 978-1-9022-3364-2

LB Discovery Bottles

ISBN 978-1-9060-2971-5

LB Music

ISBN 978-1-9041-8754-7

All available from

www.acblack.com/featherstone

The **Little Books** series consists of:

All Through the Year
Bags, Boxes & Trays
Bricks and Boxes
Celebrations
Christmas
Circle Time
Clay and Malleable Materials
Clothes and Fabrics
Colour, Shape and Number
Cooking from Stories
Cooking Together
Counting
Dance
Dance, with music CD
Discovery Bottles
Dough
50
Fine Motor Skills
Fun on a Shoestring
Games with Sounds
Growing Things
ICT
Investigations
Junk Music
Language Fun
Light and Shadow
Listening
Living Things
Look and Listen
Making Books and Cards
Making Poetry
Mark Making
Maths Activities
Maths from Stories
Maths Songs and Games
Messy Play
Music
Nursery Rhymes
Outdoor Play
Outside in All Weathers
Parachute Play
Persona Dolls
Phonics
Playground Games
Prop Boxes for Role Play
Props for Writing
Puppet Making
Puppets in Stories
Resistant Materials
Role Play
Sand and Water
Science through Art
Scissor Skills
Sewing and Weaving
Small World Play
Sound Ideas
Storyboards
Storytelling
Seasons
Time and Money
Time and Place
Treasure Baskets
Treasureboxes
Tuff Spot Activities
Washing Lines
Writing

All available from

www.acblack.com/featherstone